Practical Statistics

Practical Statistics

Robert E. Levasseur, Ph.D.

MindFire Press
Annapolis, Maryland

ISBN–13: 978–0–9789930–1–6
ISBN–10: 0–9789930–1–2

Library of Congress Control Number: 2006908870

Published by MindFire Press of Annapolis, Maryland (www.mindfirepress.com)

"The mind is a fire to be kindled,
not a vessel to be filled."

Plutarch

Table of Contents

A Note from Dr. L

This is a book about a very important subject, something we use every day, whether we realize it or not—statistics. Will it rain? Will construction on the road to work necessitate taking a new route? Will my husband remember my birthday? Will I get an A on the final? Who will win the next election? These and a myriad of other questions have answers, as we shall see, which are rooted in statistical reasoning and analysis.

According to my Merriam Webster dictionary, statistics is "a branch of mathematics dealing with the collection, analysis, and interpretation of masses of numerical data." Simply put, it is an approach to analyzing data that uses statistical reasoning.

Most statistics books, even the so-called easy ones with titles which suggest that anyone can learn statistics by reading them, do not focus on statistical reasoning and practical application. Instead, they emphasize theory and mathematical manipulation. As a result, many people trying to learn the essentials of this important subject (particularly the majority who are not strong in math) find themselves intimidated and confused, rather than engaged and informed.

The purpose of this book is to provide a pragmatic, application-focused alternative to the existing books in the field for anyone who wants to learn something about statistics, so he or she can use it to make better decisions. For this reason, it assumes only knowledge of basic arithmetic and a willingness to learn. Yet, it shows how to use mathematics to support statistical reasoning when solving important problems of testing, estimation, and forecasting.

While this book is for problem solvers and decision makers, not mathematicians, it is a real statistics book. As a professor of management who has taught statistics to students at the undergraduate and doctoral levels, I have made sure of that.

This book is an integrated treatment of the subject, leading progressively from fundamental concepts to more advanced ideas. Rather than treating them in a highly theoretical manner, or worse, leaving them out, I introduce the necessary mathematical formulas, which are exceedingly powerful if used correctly, in as intuitive and non-threatening a manner as possible.

Some other features of the book include: (a) the use of a unique, decision model framework—to build a picture of practical statistical reasoning, analysis, and decision making that cumulatively integrates theory and practice; (b) concrete examples of real-world problems with detailed explanations of their solutions—to illustrate the application of statistical concepts and mathematical formulas; and (c) optional pop quizzes with answers and explanations of the methods used and why—to help test and refine statistical reasoning and analysis.

Students interested in (or required to study) statistics should find this book useful as an introduction to the field. Therefore, they should read it before tackling a more advanced, theoretical book or textbook.

Decision makers who want to improve the quality of their problem solving should find this book's focus on decision models and statistical reasoning a comfortable, welcome alternative to the abstract theoretical focus of most management texts on the subject or the number-crunching, black–box feel of most spreadsheet tools available for statistical analysis.

Finally, interested readers who just want to learn something about statistics for their personal use should find the consistent focus on concrete, real–world examples and practical data analysis helpful and motivating.

I have tried to write a book that is straightforward and complete enough to demystify statistics while conveying the essential elements from which it derives its power. After you read it, I hope you agree.

Chapter 1

Why Study Statistics?

Why would anyone in his or her right mind want to study statistics?

There are lots of reasons. Many, for whom statistics is required in a program of study, have no choice. Some, particularly managers and professionals, are interested in learning how to make better decisions. Others, like researchers, know it will enable them to make sense of their data and present their findings to colleagues in a common language. A minority, who love mathematics, just want to learn powerful statistical methods and add them to their mathematical tool kit.

What all students of statistics whose teachers properly introduce them to its study eventually discover and come to appreciate is its great utility. As the examples in this chapter will show, statistics is just plain useful to anyone trying to make sense out of the quantitative aspects of life.

The table below contains half a dozen typical situations amenable to statistical reasoning and analysis. These include two everyday occurrences (attending a picnic or a softball game), two normal business activities (manufacturing process control and sales forecasting), and two other regular events (an election and an experiment). In this chapter, we will examine how statistics can inform and improve decision making in each situation.

Situation	Question	Statistical Focus
Picnic	Will it rain?	Probability
Ball Game	Will she get a hit?	Batting Average
Production	Is the quality high?	Confidence Interval
Forecasting	What is the demand?	Expected Sales
Election	Will he win?	Probability
Drug Test	Does it work better?	Level of Significance

Don't worry about the last column yet. You will learn the meaning of these statistical terms at the right time.

A Sunday Picnic

While the organizers of a family reunion picnic have to consider many factors, we will focus exclusively on the weather for the day of the event to keep this introduction to statistical reasoning simple. Therefore, the basic question is "Will it rain or shine?"

Because of your familiarity with the picnic site, imagine that the organizers asked you, as a family member, this question. How would you go about determining a thoughtful answer?

Here is one scenario for how you might do it.

First, knowing the location of the picnic site would give you a general idea of the likelihood of rain being a problem. For example, a picnic held in the Pacific

Northwest would stand a greater chance of being a washout than one held in Arizona.

Second, you would want to know when the organizers planned to hold the picnic. Together with the location, this information would give you a fair idea of the likelihood of rain based on the general climate in the region at that time of year.

With this local climate knowledge, you would be in a position to make an educated guess (i.e., a forecast) about the likelihood of the day of the picnic being sunny or rainy. Let's talk a little about how you might express your educated guess to the organizers.

What is the lowest number that this probability can be? By convention, the answer is zero, which means no chance of rain. What is the highest number that it can be? A certain or known outcome has a probability of one. So, the probability of rain for any given day in any given location lies between zero and one. Or, in mathematical terms, $0 \leq P(\text{Rain}) \leq 1$, where P stands for probability.

Returning to our scenario, imagine that the reunion picnic will occur in a location and at a time of year during which it rains only about 25% of the time (i.e., it rains on average about one out of every four days) according to climate data that you have accessed by doing a search on the Internet.

As a result of your search and data analysis, you tell the organizers that the probability of a rainy day for the reunion picnic is 0.25. Based on your estimate, they decide to hold the picnic on the planned date.

A Note on Probability

Before moving on, let us express this estimate in different, but equivalent, terms to make sure you understand this fundamental notion of probability.

Here are five equivalent ways to express your estimate:

- P(Rain) = 0.25
- P(Shine) = 0.75
- The chance of rain is 25%.
- The chance of a sunny day is 75%.
- The odds of a rainy day are 1 to 3.
- The odds of a sunny day are 3 to 1.

Note that the odds of a rainy versus sunny day are 1 to 3, not 1 to 4. Why?

To answer this question, consider this hypothetical weather data for the day and location of the picnic for the past dozen years. Note that R stands for a rainy day and S for a sunny day.

R S S R S S S S R S S S

There were three rainy days and nine sunny days during this period; or, one rainy day for every three sunny days. Hence, the odds of the next day in the series being a rainy day are 1 to 3.

Alternatively, the probability of a rainy day based on this data is:

P(Rain) = Number of Rainy Days
          Total Number of Days

P(Rain) = 3/12 = ¼ = 0.25 = 25%

The probability of the next day in the series being a rainy day is 1 in 4. Hence, odds and probability are related, but they are not the same.

Pop Quiz

Question: If the odds of it raining on a given day are 2 to 3, what is the probability that it will rain?

Answer: The probability of rain is 0.4 or 40%.

Explanation: According to the odds, it rains two out of every five days. Hence, the probability that it will rain is two divided by 5 or 40%. Expressed mathematically, P(Rain) = 2/5 = 0.4.

If you don't understand these concepts yet, don't get discouraged. Just keep reading. Soon you will begin to understand at least the nature of probability and statistics. Later you can fill in the theoretical and mathematical blanks.

A Softball Game

It's the proverbial bottom of the ninth. The bases are loaded and the score tied. Lacey, the home town hero, is at the bat. Will she get a hit or not?

Because you've learned something about probability and statistics from your work on the reunion picnic weather problem, you know that the statistical question in this case is, "What is the probability that Lacey will get a hit?"

Your first thought is that her batting average is the place to begin. Lacey is a 0.400 hitter, which means that she gets a hit 40% of the time, or 4 out of every ten times at bat. Without any further information, you would say that the probability of her getting a hit is 0.4.

Your second thought is that Lacey is a terrific hitter with runners on base. In fact, her batting average with one or more runners on is 0.500.

Your final thought is that Lacey has trouble hitting lefties. Knowing this, the opposing team's manager has brought in a left-handed reliever from the bull pen to face Lacey. Lacey's batting average against lefties is 0.300.

So what is the probability that Lacey will get a hit? Is it 0.3 (her average against lefties), 0.4 (her overall average), or 0.5 (her average with runners on base)?

There is no exact way to determine the probability that Lacey will get a hit. But my guess would be that her average with runners on and her average against left-handers would balance each other out. So, I would average these two averages to get an estimate of 0.4— i.e., (0.3 + 0.5)/2.

Coincidentally, 0.400 is her overall batting average, which makes me feel even better about my estimate. This is because, without knowledge of her average with runners on or against lefties, the best estimate would be Lacey's overall batting average. From a statistical standpoint, anything else would be a hunch.

Now, this is a somewhat artificial situation because only a statistician would carry around data on batting averages under different conditions like we needed to answer this question; but the reasoning we used to determine an answer is sound, with numbers or without. To illustrate, imagine a conversation between two savvy home town fans going something like this:

"The bases are loaded, and Lacey eats these situations up. She'll knock in the winning run."

"But they're bringing in a lefty, and Lacey struggles against lefties."

"Yeah, you're right, but I still think she can do it. After all, she's a four hundred hitter. So, she has almost a fifty-fifty chance of getting a hit."

Clearly, our use of statistical formulas to manipulate the numbers adds concreteness to these estimates, but it does not change the underlying (sound) statistical reasoning. As you will see, this central theme of sound reasoning supported by sound analysis will resonate throughout the book.

A Production Process

Imagine that you are a production supervisor for a consumer products company that sells a popular brand of cereal. If each box must contain a certain amount of cereal, W as measured by weight, how do you monitor the production process to ensure that W, no more and no less, ends up in each box?

The answer is with statistical quality control methods.

Okay, let's put aside the technical jargon for a minute and think about the problem. Why does your company want you to fill each box with exactly W (say 20 ounces of) cereal? To see why, answer these two related sets of questions for yourself:

- Who would care if you filled each box with less than W? Why?

- Who would care if you filled each box with more than W? Why?

You have probably figured out that the "who" in the first case are consumers of the cereal, who want their moneys' worth; and that the "who" in the second case is the company, which doesn't want to give away extra cereal for free, because they are in business to make a profit.

But how, you ask, can you be sure of getting exactly W in each box? Actually, you can't, which is why you, as the production supervisor, need statistics to solve this problem.

In practice, given the imperfect nature of all production processes, the best you can hope for is to get the weight of an average box (W average) to fall within a reasonable interval that includes W as its central value, such as W plus or minus 1%.

Given this hypothetical range set by your management, your goal as a supervisor would be to run a production process that fills your 20 oz. boxes with 19.8 to 20.2 ounces of cereal. [Note that 1% of 20 is 0.2, and 20 ± 0.2 = 19.8 to 20.2.]

So, how would you use this information to ensure that the proper amount of cereal found its way into every box?

The answer is that you would take samples of boxes at random from a production run, and weigh them. Then you would compute the average weight of all of the boxes from your sample, and compare that average weight to the desired range of weight—19.8 to 20.2 ounces in this case. If it fell inside the range, you would be reasonably sure that the process was "under control" in the sense that it was producing boxes filled with the right amount of cereal (W). If, however, the average weight of the sample of boxes fell above or below the range, you would suspect that you had a problem. If this happened several times in a row, it would be clear that the process was "out of control" and that corrective action was necessary.

This procedure, called statistical quality control, uses statistics to control the quality of the final product, be it a widget or a box of cereal. It is a standard, easily implemented technique used in plants all over the world to control manufacturing processes.

A Note on Sampling

Although the purpose of the material in this chapter is primarily to provide you with insights into statistics and a sense of how useful it is, I don't want to miss an opportunity to provide those of you who are interested in the supporting mathematics with additional information.

So, here is an initial primer on sampling. If it proves to be too much for you, just move on to the next section on sales forecasting and come back to it later when you feel more knowledgeable about statistics. Remember, learning statistics, like any challenging subject, is an iterative process. So, don't be discouraged if you don't understand something the first time you read it. Just be persistent in reviewing it, and you will eventually figure it out.

Meanwhile, let's get back to sampling theory and your boxes of cereal.

Let's perform a few simple thought experiments to help us understand the nature of sampling. We begin by selecting an entire production run as our sample. That is one hundred percent of the population of boxes from which we could have selected a smaller sample if we had chosen to do so. Let's call the number of boxes in this entire run, N.

If we compute the average weight of the boxes in this super sample of all boxes in a run, we will have determined the exact average weight of the boxes in this population, without any error in our estimate. Say that we do this for an entire run and find that the average weight is precisely 20 ounces, our ideal target weight. This makes us happy as it suggests that the process is working well.

As you can see, if we do 100% sampling of all the boxes we produce, we can exactly determine the weight of the average box and exactly determine if it lies in the predetermined range (from 19.8 to 20.2 in this case).

Does this seem like the right thing to do? Why? Why not?

Well, if we decide to weigh every box, the company will incur a significant cost that it will have to pass on to the consumers in the form of higher prices (which could lead to decreased sales in the highly competitive cereal market) or eat the extra cost and, thereby, erode its profit margins. Neither of these alternatives is likely to get you the promotion to manager that you want.

So, what do you suggest we do? The answer is to select a sample—a smaller number of boxes (n) than the total number (N) produced in a given run. Thus, our sample consists of n boxes, where n < N.

Weighing this smaller sample will cost the company less, but create a problem which we will have to figure out how to manage. Namely, we will not know the exact average weight in our population (a production run of size N) based on the average weight of the boxes in our sample on n items. In short, we will have to deal with some uncertainty in our estimate of the population average (also called the population mean) based on our estimate of the sample average (also called the sample mean).

This is where statistics comes into play. Up to now, we have been working with simple arithmetic averages of the weight of a number of boxes. However, we must now determine the probability that a given sample average (based on n boxes) is an accurate estimate of the population mean (based on all N boxes).

To illustrate the problem, imagine that we have a very small hypothetical production run of ten items with the following weights (rounded to the nearest unit to keep the arithmetic simple):

16, 20, 19, 21, 20, 20, 21, 22, 20, 21

What is the population mean? Adding up the weight of the ten boxes and dividing by ten gives us a mean of exactly 20.

What is the sample mean? Well, that depends on the size of the sample?

For example, if you chose a sample size of one (n = 1) and weighed the first box off the line, the sample mean would be 16. But, if your sample size was two (n = 2) and you weighed the first and second boxes off the line, the sample mean would be 18 [i.e., (16 + 20)/2]. What if your sample size was four and you selected the first four boxes as your sample. In that case, the sample mean would be 19 [i.e., (16+20+19+21)/4].

The following table summarizes the results of our hypothetical sampling schemes:

Sample Size	Sample Mean	Population Mean
1	16	20
2	18	20
4	19	20
10	20	20

From the table, it is clear that as sample size increases, the sample mean becomes a better estimate of the population mean. This finding is not a coincidence. It illustrates a very important law of statistics, called the Central Limit Theorem, which you will learn more about later.

A Sales Forecast

If you were responsible for forecasting your company's sales for the coming year, how would you do it? Before you answer that, let's talk a little bit about forecasting.

Case A

If the data below were the annual sales (in millions of dollars) for the last five years, what would your forecast be, and why?

10, 10, 10, 10, 10, **?**

I bet you would say $10 million, all other things being equal, because the company has made that amount each year for the last five years.

Case B

But, what would you forecast if historical sales were as follows:

10, 12, 8, 11, 9, **?**

Not so clear is it? Statistically, your best estimate of sales would be the historical average or mean sales of $10 million—i.e., [(10+12+8+11+9)/5].

Case C

What would you forecast if these were the annual sales:

10, 11, 12, 13, 14, ?

My guess is that you would say $15 million, since the trend during this period has been an increase in sales of $1 million per year.

Case D

But, what would your estimate of sales be if this were the historical sales data:

10, 12, 11, 14, 13, ?

As in Case B, where the meaning of the data is also not self-evident, what to forecast for next year's sales is not clear. There is an upward trend in the data, but how do you take it into account?

By using an advanced statistical analysis method known as regression analysis, we can develop a model or mathematical equation that we can use to forecast sales in this case. Leaving the mechanics until a later time, we focus on the formula (given below), and how to use it to compute the estimated sales for next year, i.e., the forecast.

$$\text{Sales} = A + B \cdot \text{TIME} = 9.6 + (0.8) \cdot \text{TIME}$$

Note that TIME varies from 1 (for the oldest data point) to 5 (for the newest data point—last year's sales).

Now, the independent variable (TIME) increases by 1 each year. Therefore, TIME = 6 for next year's sales. Substituting 6 for TIME into the equation above and doing the math yields a sales forecast for next year of $14.4 million. [As an exercise, I encourage you to run the numbers with a calculator and convince yourself that 14.4 is indeed the forecast of the regression model.]

If you were actually responsible for forecasting your company's sales for the coming year, knowing that your estimate would serve as the basis for major marketing, financial, and operational decisions, would you rely on your gut feel alone, or would you augment your intuitive sales estimate with the results of a sophisticated statistical technique, like regression analysis?

Early in my business career, I worked for a Fortune 50 consumer products company as a marketing analyst. My job was to help the brand managers make these types of forecasts for their major products.

Based on this experience, I can assure you that most managers would opt for the statistical estimate and use their judgment to modify it to incorporate factors not included in history which might have a significant effect on the forecast, such as the anticipated emergence of a new competitive product in the market place.

Being a savvy manager, I imagine that you would do the same thing when forecasting your company's sales.

An Election

Who will win the election, Candidate A or Candidate B?

Suppose that Candidate A hired you to determine his chances of becoming mayor of your small town. How would you go about doing this?

One way would be to ask every one of the 10,000 voters in town which of the two candidates he or she planned to vote for on Election Day. While this would certainly give you a very good idea, subject to people changing their minds at the last minute, of who would win, it would be very expensive to track down every voter and get a yes, no, or undecided from each.

Another way would be to ask a carefully selected subset of voters—a sample—for whom they plan to vote. This choice, while more cost effective than 100% sampling (which is called a census), also has some pitfalls that you must deal with. For example, if you were to ask say 100 people for their opinion, would that be enough to get a decent answer? Also, would the answer you got depend on which 100 people you asked?

Let's deal with the second question first. Does it matter whom you sample? Before you answer, think what the likely answer would be if you asked 100 supporters of either candidate. Clearly, the quality of your estimate does depend on the people you ask. So, how do you ensure that you ask the right people? You do it by drawing a representative sample, at random, from all of the possible choices (voters in this case).

By representative, we mean a sample that has roughly the same characteristics as the total population of voters in your town. Choosing them at random ensures that you do not introduce any selection bias by intentionally picking certain types of people.

In your case, obtaining an up-to-date roll of registered voters in alphabetical order might give you a pretty good list to pick your sample from. Because you need 100 out of the 10,000 voters on the list, you must select every 100th voter for your sample (i.e., 100 • 100 = 10,000). To make this a random selection, you simply need to pick the first voter at random from one of the first 100, then you pick every 100th voter thereafter.

If you did not have a random number table handy (who but a serious student of statistics has one), you might ask someone with no knowledge of the list to choose a number from 1 to 100. If the number were 43, you would select the 43rd, 143rd, 243rd, and so forth names off of the list to get your sample of 100 people.

You would then ask each of these 100 people whom they planned to vote for in the upcoming election, and compute the percentage for Candidate A or Candidate B or undecided.

Let's say you did that, tabulated the votes from your informal poll, and discovered that 52% favored A, 43% favored B, and 5% were still undecided. What would this mean? That A is the clear favorite? It certainly looks that way, since even if the undecided voters voted for B, A would still have a winning percentage.

Perhaps, but you would be leaving something extremely important out of your analysis of the results of the poll; namely, the degree of uncertainty that a poll of a mere 1% of the population of registered voters in the town accurately measures the preferences of all 100% of them.

Another Note on Sampling

If you get the point of this section—that election polling depends on statistical sampling theory for its accuracy—and don't want to learn any more about sampling at this time, simply skip forward to the drug testing situation. If however, as I hope, you are curious enough to read on, please do so and learn more.

We now know that the quality of your estimate depends on the way that you go about selecting the sample. Therefore, in this note, we will focus on the other question posed earlier. "If you were to ask say 100 people for their opinion, would that be enough to get a decent answer?"

At its core, this is a question about uncertainty. All other things being equal, if you ask one person, you have the maximum uncertainty in your estimate. If you ask everyone in the population, you have the minimum uncertainty. If you ask more than one person, but less then all of them, the degree of uncertainty will fall somewhere in between. Let the number you decide to ask be n, where $1 \leq n \leq N$. Note that N is 10,000 in this hypothetical case.

To make this clear, consider what would happen if all but one person intended to vote for B, but you asked this one person for her preference. You would predict a victory by A on election day. Wouldn't you feel foolish the day after the election? Conversely, if you asked all but one of the voters, not asking the last one how he or she might vote would not affect your estimate by any measurable amount.

So, uncertainty goes down as sample size (n) goes up. Knowing the exact formulas that relate the allowable error in the forecast to sample size would permit you to determine if a sample of 100 voters would be enough or if you would need a larger sample to get the accuracy required in your forecast.

The next time we have a presidential or other major election for which they conduct and report on exit polls, listen for both the percentage of the vote predicted for each of the major candidates and a statement about the range of accuracy of the poll, such as "This poll is accurate to within plus or minus X percentage points." For example, if candidate A has a predicted percentage of 52% and the exit poll is accurate to within 3 percentage points, then it is likely that the true percentage for A will fall in the range from 49% to 55% (i.e., 52 ± 3).

Note that it is not a certainty that Candidate A's final, tabulated percentage of the vote will be between 49% and 52%, it is just very likely. It could, in fact, statistically speaking be much more or less with a small, but finite probability. Why this is so is something that we will explore in the next section.

An Experiment

When a company tests an advertising campaign, it hopes to discover the superiority of the new message versus the old. When a pharmaceutical company tests a new drug, it hopes to show its efficacy in comparison to existing drugs. When a social science researcher tests the effectiveness of transformational leadership, she hopes to demonstrate its power to move mountains when compared to traditional, top-down leadership styles.

Is it possible in each of these cases to "prove" conclusively the benefits of the new versus the old based on a well designed experiment? If so, why? If not, why not?

Avoiding a philosophical discussion about the nature of proof (i.e., whether it is really possible to prove anything), we examine a practical argument for the meaning of proof in experimental settings. In so doing, we seek to demonstrate how probability enters into the decision of whether an experiment was successful or not.

How likely is it that you will win the lottery if you buy a single ticket? Not very, of course. Nevertheless, you know that, however remote your chances, there is a finite probability of winning, or you would not have purchased the ticket in the first place. After all, stories abound of people who have bought a single ticket and won a lottery worth millions.

Given this insight into the nature of probability, let us consider the design of a simple experiment. The basic idea is to randomly assign participants to test and control groups. The test group will receive the new, experimental treatment—the new ad, the new drug, transformational leadership, etc. The control group will receive the old method—the old ad, the old drug, traditional leadership, etc.

Then, the researcher will measure the effect of the new treatment on each of the members of the test group and calculate an average or mean effect for the test group; and, likewise for the control group.

The goal of the experiment is to test the belief (called a hypothesis) that the new method is better than the old one. By comparing the two group means statistically, it is possible to test this hypothesis with a high degree of certainty, but not perfect certainty.

For example, there may only be 1 chance in 100 (i.e., a probability of 0.01) of the difference between the mean test group effect and the mean control group effect being what it turned out to be. But there is still that chance.

Hence, the researcher will conclude with a high level of confidence (99%) that the test provided confirmation of the hypothesis—that the experimental treatment is better than the old method—but she will also know that this experiment does not provide absolute proof that the new method is superior to the old (because of that pesky 1% chance of being wrong).

Conclusion

Hopefully, as a result of working your way through the situations presented in this chapter, you are starting to appreciate the power and usefulness of statistics. In the next chapter, we will examine a simple model that will provide you with a framework for using statistics to systematically analyze problems involving uncertainty, like the ones inherent in the six situations discussed in this chapter.

Chapter 2

A Simple Statistical Decision Model

Kurt Lewin, the renowned humanistic psychologist, once said, "There is nothing so practical as a good theory" (Marrow, 1969, p. viii). A model is a way of representing a theory, whether it is good or bad, in the form of a framework or structure that incorporates its main variables.

In *The Human Side of Enterprise*, Douglas McGregor presented two theories of motivation, which he called Theory X and Theory Y. These two theories represent opposite ways of looking at people's motivation to work.

For our purposes, we focus on Theory Y, which basically says that because people are inherently motivated to do good work, treating them as if they want to do a good job, e.g., by giving them encouragement and support but not micromanaging them, will motivate them to do their best (McGregor, 1960).

Expressed as a model, Theory Y looks something like this:

Positive Treatment ⇨ Good Work

In short, assuming that people want to do good work and treating them accordingly motivates them to do good work.

In the previous chapter, we operated on the basis of an underlying theory that relevant statistics, properly analyzed, would enhance decision making. For example, knowing the probability of rain would improve the quality of the decision to hold the picnic on a given day. Similarly, knowing Lacey's batting average would improve the quality of our estimate of the likelihood that she would get a game-winning hit. And so forth.

Expressed as a model, that underlying theory looks something like this:

Statistics ⇨ Decision Model ⇨ Decision

In other words, this model encapsulates the theory that relevant statistics, when properly analyzed (by means of an appropriate decision model), result in better decisions.

As we shall see, this simple statistical decision model, and the more elaborate variations of it presented later, will serve as a framework for determining how we can use statistics to enhance decision making in a wide array of practical situations.

Chapter 3

Data in Statistical Decision Making

Data plays a key role in statistical decision making. In fact, it is the source of our statistical information as shown in the following enhancement of the basic model:

Data ⇨ Statistics ⇨ Decision Model ⇨ Decision

Given the pivotal role of data in the analytical process, it is vital that we know the various forms of data and their applications and limitations.

To this end, we will first examine two distinct schemes used to classify data—by type and by level of measurement. Then, we will integrate them into a coherent picture.

Types of Data

According to Berenson, Levine, and Krehbiel (2004), there are two types of data—categorical and numerical. Furthermore, numerical data comes in two forms— discrete and continuous. What do these terms mean?

As the respective words suggest, numerical relates to numbers, and categorical is about categories. Hence, *if you can do arithmetic operations on it, it is numerical data. If you can't, it's categorical data.* Here are several examples of each type (one data set per line). See if you can tell which are numerical and which are categorical.

- 1, 2, 3, 4

- A, B, C, D

- 1.14, 2.28, 4.32, 6.46

- Male, Female

If you deduced that the odd lines have numerical data and the even ones categorical, you are right. To illustrate, you could add the four numbers in the first set of data and get 10 (i.e., 1+2+3+4 = 10), but adding four letters in the second data set wouldn't make sense because A+B+C+D doesn't mean anything. Similarly, you could add the four numbers in the third example and get a meaningful total (14.2), but Male plus Female is a meaningless entity.

Now that you know the difference between numerical and categorical data, you need to learn the difference between discrete and continuous numerical data.

This too is simple. *If you can count it, it's discrete. If you can't, it's continuous.* For example, if I asked you how many brothers and sisters you have, you could count them and announce the result. But, if I asked you how tall you were, you might say 5′ 10″ or 6′ 3″ or any other number from very small to very tall, because people don't come in discrete sizes. In other words, height is a continuous variable which results from a measuring process, not a discrete one resulting from a counting process.

Pop Quiz

Question: Which of the two numerical data sets on the previous page is probably discrete, and which is definitely continuous?

Answer: The first is probably discrete. The second is definitely continuous.

Explanation: You might have 1, 2, 3, 4 (or more or fewer) brothers and sisters, but you can't have 1.14, 2.28, 4.32, or 6.46 of them. Also, 1, 2, 3, and 4 may be the result of a counting process, but they may also be elements in a larger, continuous data set, such as 1, 1.14, 2, 2.28, 3, 4, 4.32, and 6.46. Hence, we use the word "probably" to qualify the answer.

Levels of Data

There are four levels of data—nominal, ordinal, interval, and ratio. The amount of statistical analysis that you can legitimately do on data varies from very little with nominal data to the maximum possible with ratio data. Hence, you must determine the level of your data to know which statistics you can legitimately use to analyze it.

Nominal means in name only. Hence, whether a person is Male or Female is nominal data.

Ordinal indicates order or rank. Hence, whether a person earns an A, B, C, D, or F in a course is ordinal data. [Why isn't Male/Female ordinal data?]

Interval focuses on the space between numbers. Hence, differences between interval data have meaning, but not their ratios. Temperature, standardized test scores, and numerical questionnaire ratings are examples of interval data. For example, it is correct to say that a temperature of 80 degrees Fahrenheit is 40 degrees higher than a temperature of 40 degrees, but it is not correct to say that 80 degrees is twice as warm as 40 degrees.

Ratio means quotient. We measure ratio data on a scale that has a true zero. Hence, dividing ratio data yields meaningful answers. For example, if you weigh 200 pounds and your best friend weighs 100 pounds, you weigh twice as much. Similarly, if you are 30 and your best friend is 20, you are 50% older than your friend.

The fact that we measure weight and age on a numerical scale which has a true zero point (i.e., for which there is such a thing as zero weight and age) allows us to perform the widest variety of mathematical operations on these data (including division) of any level of data.

Remember, the level of your data determines which statistics you can apply to it. The closer your data are to ratio data (on the nominal, ordinal, interval, ratio hierarchy), the more mathematical analysis you can perform on it. Conversely, the closer you data are to nominal data, the less you can legitimately do with it. While you will learn even more about this later, the example provided after the composite view, which involves the analysis of survey data, will serve as a brief introduction to the subject.

Pop Quiz

Question: In conducting a survey, you capture the class of each student who participates because you believe the responses will vary by class. What level of data is the set of categories, Freshman, Sophomore, Junior, or Senior?

Answer: Participants' class is ordinal data.

Explanation: Although Freshman, Sophomore, Junior, and Senior are names, they are also an ordered set from lowest to highest class. Hence, they are ordinal data.

A Composite View

If we put these two data classifications together, what do we get?

First, categorical data is either nominal (e.g., Male, Female) or ordinal (e.g., A, B, C, D).

Second, numerical data is simultaneously discrete or continuous and interval or ratio, as the examples in the table below show.

Type\Level	Interval	Ratio
Discrete	Ratings (1 to 5)	Books Owned
Continuous	Temperature	Height, Weight

A Practical Note on Data and Analysis

For the sake of discussion, imagine that you have conducted a survey of students in your school to determine how often they use the computers in the library. Your objective is to provide the administration with data and statistics for deciding whether there is a need more or fewer computers and whether they should purchase PCs or Macs.

Among other things, you have data on (a) participant class, (b) frequency, time of day, and duration of usage of the computers, (c) reasons for using or not using the library computers, and (d) preferences for PCs or Macs. How do you go about analyzing the data?

First, participant class is categorical ordinal data. Therefore, computing a statistic like the average class of the participants does not make sense. However, you could segregate your numerical data by class, analyze it for each class, and then aggregate it into a total picture for the entire school (assuming, of course, that you took a representative, random sample of students from the four classes). You could also determine, by means of appropriate statistical methods (e.g., analysis of variance or multiple regression analysis) if there were a significant difference among the classes in terms of frequency, time, and duration of usage, or preference for PCs versus Macs.

Second, usage data is numerical ratio data. How often, when, and how long questions generate numerical answers such as once, twice, or five times a week, 8:00 am or 4:00 pm, and 10 to 20 minutes, respectively. So, calculating an average, such as 3.65 times per week or 28 minutes per session, makes sense for these data. However, because frequency is discrete data, you might express your average as 3 to 4 times per week, rather than 3.65, to not suggest that there is such a thing as using a computer 3.65 times a week.

Note that you can also compute an average time of day, although it may not be a very meaningful statistic, by first converting the times to a 24 hour clock. When measuring time this way, add 12 to any time after noon to differentiate it from the first 12 hours of the day. Hence, 1 am is 1:00, but 1 pm is 13:00, and 6 am is 6:00, but 6 pm is 18:00.

So, although it is permissible to compute an average time of day (after all, this is ratio data), it is not very helpful to do so. A better idea would be to present a graph of usage as a function of the time of day. Then, the peak periods and idle times, which an average would mask, would be evident.

Third, reasons for using the library computer is categorical nominal data, because it contains no numbers. However, this does not mean that you cannot provide statistical information based on your analysis of this non-numerical data. How might you do this?

If you're ready, here is the answer. If you take note of how many times each reason is given, you can rank order the list of reasons from most to least frequently mentioned. This is a very helpful way to present qualitative data in that it gives the reader a sense of the importance of each item based on the percentage of the participants in the survey who mentioned it. In fact, it would be a good idea to compute and provide this numerical information along with the reasons in a table like the one below.

%	**Reason for Using or Not Using**
70	Always being used by other students.
50	Only use Macs.
20	Too busy to use.
10	I don't own a personal computer.
5	They're conveniently located.

This data suggests that if the school purchased more personal computers, particularly Macs, to complement the existing PCs in the library, the students would probably use them.

Finally, preference data is numerical discrete interval data because the participants rated their preferences on the following (simplified) scale:

Prefer PCs	**Indifferent**	**Prefer Macs**
1	2	3

Note that while the individual participant ratings are interval data, the average by class and the overall sample average is ratio data. [Why?]

The reason becomes clear if we examine a simple case. Consider the ratings of five students on this question—3, 1, 3, 3, 3. This is clearly discrete data. But, if we average the ratings we get 2.6, which is definitely continuous.

This statistic (i.e., the average preference) suggests a strong preference for Macs, which corroborates the findings of the previous analysis of the categorical data on the reasons for using or not using the computers in the library.

Conclusion

Now you know that the type and level of your data directly affect the statistical analysis you can perform on it, which in turn affects the information available for decision making.

If your data already exists, then you need to determine what you can do to make statistical information out of it. Whether this meets the requirements of your decision model or not may be problematic.

However, if you have the option of collecting your own data, use the statistical decision model in reverse to guide your actions. That is, based on the decision you want to make, determine the nature of the decision model. Then, determine the statistics needed to use that decision model to make the decision. Finally, determine the level and type of data you will need to compute those statistics.

Chapter 4

A Primer on Descriptive Statistics

The purpose of this chapter is to provide you with an overview of statistics—a big picture view—so you won't get lost in the details. Up to now, we have taken a less systematic approach, preferring instead to focus on applications first and supporting statistical reasoning and analysis second. Now, we tackle the fundamentals of statistical theory head on, but from a high–level, macro perspective.

The model that will guide us in our explorations is an expansion on the first part of the model introduced in the last chapter, which itself is an elaboration on the simple statistical model presented in chapter 2. Namely,

> Data ⇨ Statistical Model ⇨ Statistics

A statistical model, for our purposes, is a set of assumptions and corresponding mathematical formulas used to turn data into information (i.e., statistics). A statistic is a single number that in some fashion describes a data set.

A Clarifying Example

A simple example should clarify what this portion of the statistical decision making model represents.

A. Here is a data set:

3, 10, 8, 2, 5, 16, 12

B. Here is a statistical model:

$$\text{Average} = \frac{\text{Sum of All Data Points}}{\text{Number of Data Points}}$$

C. Here is the resulting statistic:

$$\text{Average} = \frac{3 + 10 + 8 + 2 + 5 + 16 + 12}{7} = 8$$

Statistical Categories

Basic statistics fall into two broad categories:

- Descriptive Statistics
- Inferential Statistics

To understand each type, we need a few definitions. First, we need one for population, which is the universe of all possible outcomes. Second, we need one for a sample, which is a subset selected from the population, often at random.

For example, if 95, 92, and 83 represent the mid-term grades of all three students in a tutorial group (i.e., the population), a sample from that population might consist of the grades of one or two of them, but not all three.

Pop Quiz

Question: What are all of the samples of one and all of the samples of two that you might select from the population of three students' mid-term grades?

Answer: Samples of size one are 95, 92, or 83. Samples of size two are 95 and 92, 95 and 83, or 92 and 83.

Descriptive statistics, as the name implies, tell us something about the data set itself, whether it is a sample or the population. That is, they describe the characteristics of the data. Inferential statistics, as the name implies, allow us to make inferences (i.e., educated guesses) about a population based on sample data.

For instance, if I gave you the following mid-term scores—95 and 83—and told you they were from a sample of two students, could you describe the sample data set using at least one statistic you have learned so far? Of course you could, if you computed the average score, which is 89 [(95+83)/2].

Now, if I asked you for your best guess about the mean (i.e., average) score achieved by all of the students who took the mid-term (i.e., the population), you would, and should, say 89, which is the sample mean you just computed. By providing this estimate, you have made an inference about the population mean based on the sample mean.

Because you know how to distinguish the central notion of descriptive statistics from that of inferential statistics, we can now explore each in more depth without getting lost or confused.

Descriptive statistics, to repeat, tell us something about the data set itself, whether the data is a sample or the entire population. Descriptive statistics consist of the mean, median, and mode (which measure the central tendency of the data); the range, standard deviation, and variance (which measure the variation of the data around the mean); and various measures of the shape of the distribution of the data.

Inferential statistics, as the name implies, allow us to make inferences about a population based on sample data. These inferences include, among others, estimates of the probability of random variables, confidence intervals based on sampling (including estimates of process quality control), hypothesis tests, and predictions based on time series or regression analysis.

I imagine that you are wondering what these statistical terms mean. I encourage you to remain patient. You will learn something about all of them before we're through. I mention them now just for the purpose of classification.

Now, we will turn our attention to descriptive statistics. In the next chapter, we will study inferential statistics.

Overview of Descriptive Statistics

If someone asked you to describe a data set statistically, they would expect you to provide them with a number of measures of the data (i.e., statistics) in each of several categories. These would include:

- Measures of the central tendency of the data
- Measures of the variation/dispersion in the data
- Measures of the shape of the data distribution

Let's use the following data set to study the statistics in each category:

12, 11, 24, 12, 13, 10, 12, 13, 12, 11

First, however, we arrange the data into an ordered array from the lowest to the highest value. This will help us to see patterns in the data more easily. [You should compare the two data sets number by number to convince yourself that they contain the same data, just in a different order.]

10, 11, 11, 12, 12, 12, 12, 13, 13, 24

Measures of Central Tendency

There are three key measures of central tendency—the mean, median, and mode. Each provides an estimate of the middle or center of the data. In the case of our ordered array of data, it is clear by inspection that the middle is somewhere around 12 or 13. But what is it exactly?

Mean

The mean or average is, as we have seen, the sum of all of the data points divided by the number of data points in the set. As we have seen, the statistical model used to calculate the mean is:

$$\text{Mean} = \frac{\text{Sum of All Data Points}}{\text{Number of Data Points}} = \frac{130}{10} = 13$$

Median

The median is the number below which (and above which) lie half of the numbers in the data set. It is the 50% percentile of the data in the sense that 50% of the data lie below and 50% of the data lie above it.

For example, if the data set consisted of an odd number of data points–such as 1, 2, 3—the median would be one of the numbers in the set. For this simple, three-number data set, the median is 2, because half of the numbers lie below it and half lie above it.

Pop Quiz

Question: What is the median of this data set— 3,1,2,5,4?

Answer: The median is 3.

Explanation: First, arrange the data into an ordered array—1,2,3,4,5. By inspection, two numbers (1 and 2) lie below 3 and two numbers (4 and 5) lie above 3.

Now, let's get back to our original, ten-point data set. Because there are ten data points, five must lie below the median and five must lie above the median, by definition. But, you say, there are only ten points (an even number), which means that all of them fall either below or above, but not on the median, in contrast to the previous examples. So, what is the median in this case?

Well, as you can see from the data below, the last of the first five data points in the ordered array is 12. Similarly, the first of the last five data points in that array is 12. Hence, the median is 12.

10, 11, 11, 12, **12,** | **12**, 12, 13, 13, 24

There is, however, one other possibility we need to cover; namely, how to compute the median when you have an even number of data points and the two on each side of the median are not the same. The following modification of our original data set illustrates the problem:

10, 11, 11, 12, **12,** | **13**, 13, 14, 14, 20

If you do the math, you will find that this ten-point data set also has a mean of 13; but its median is not 12, but some number between 12 and 13. By convention, the mid-point between the two is considered the median in cases like this. Therefore, (12+13)/2, or 12.5, is the median.

Now that you know the first two measures of central tendency, the mean and the median, you are ready to learn about the mode.

Mode

The most frequently occurring value in the data set is the mode. For example, in the data set—1, 2, 2, 3—the number 2 occurs most often (twice) and is, therefore, the mode. In the data set 1, 2, 3, there is no mode, because all of the data occur with the same frequency (i.e., once).

Pop Quiz

Question: What is the mode of the original, ten-point data set—10, 11, 11, 12, 12, 12, 12, 13, 13, 24?

Answer: The mode is 12.

Explanation: Here is a list of the numbers in the data set, and, in parentheses, their frequency of occurrence. Clearly, the mode is 12, because it appears the most (four) times—10 (1), 11 (2), 12 (4), 13 (2), 24 (1).

A Note on Estimating Central Tendency

If you had to estimate of the central tendency of a data set, would you use the mean, median, or mode? Why?

Let's examine three data sets to see why there is no pat answer to this question.

Data Set One: 1, 2, 3, 3, 6

What are the mean, median, and mode of this data? The mean is 3; the median is 3; and the mode is 3. So, in this case, your estimate of central tendency is 3.

Data Set Two: 1, 2, 2, 5, 5

What are the mean, median, and mode of this data?

The mean is 3; the median is 2; and the modes are 2 and 5. So, in this case, what you should provide as an estimate of central tendency is not so clear cut? Perhaps the mean of 3 would best represent the middle of the data.

Data Set Three: 1, 2, 3, 3, 16

What are the mean, median, and mode of this data?

The mean is 5; but the median and the mode are 3. In this case, it appears that the fifth number, which is so much greater than the rest, has a significant impact on the mean. So much so, that in this case the median (and mode) of 3 would probably be the best estimate of the central tendency of the data.

In fact, in all cases where a few data points are far apart from the rest of the data, the median or mode (because outlying data points do not affect them) may provide better estimates of central tendency than the mean. [To prove this point to yourself, change the last number in the data set above to another large number (relative to the first four in the data set), say 26, and determine the mean, median, and mode of the data. Only the mean will have changed.]

This is why, when asked to provide a measure of the central tendency of a data set, the best course of action, particularly if the set contains a large number of data points, is to provide the mean, median, and mode, along with an explanation of how and why you chose one of them as your estimate.

In the last case, you might say something like this, "Because of the potential outlier in the data (16, the largest value), the mean (5) is most likely an overestimate of the central tendency in the data. Therefore, the median/mean of 3 is a better measure of the middle or center of the data."

Although we cannot go into it here, there are statistical methods for examining so-called outliers, like the number 16 in our simple example, to determine if it is okay to exclude them from a data set or not. Short of performing such an analysis, your best recourse when confronted with a potential outlier is to provide all three measures of central tendency and a rationale for your choice of one of them as your estimate.

Statistical Models of Central Tendency

Before moving on to the measures of dispersion in a data set, let's summarize what we have learned about measures of central tendency by classifying the mean, median, and mode in terms of the statistical model used to compute each.

Statistical Model	Statistic
Average of data points	Mean
Mid-point of data points	Median
Most frequent data point	Mode

Measures of Variation

We have studied ways to estimate the central tendency of a set of data. Now, we examine measures of how the data varies around that central point. The principal measures of variation or dispersion are the range, standard deviation, and variance.

We will use the same data set, arranged in an ordered array, to study the statistics in each category:

10, 11, 11, 12, 12, 12, 12, 13, 13, 24

For the center of the data, we will use the mean, which is the standard used in the computation of variance and standard deviation.

Range

The difference between the largest number in the data set and the smallest number is the range of the data. [What is the range of the data above? The answer is 14, which is 24 − 10. Note that it is correct to say that "the data ranges from 10 to 24," but it is not correct to say that "the range is 10 to 24." The range is 14.]

Standard Deviation

In standard deviation, we encounter for the first time a statistic that requires more elaborate mathematics, although it is still basic arithmetic, and is, therefore, somewhat more difficult to understand. That said, we cannot overestimate the importance of standard deviation, which has many practical uses in statistical analysis. Hence, you must learn what it is and how to compute and use it.

Standard deviation is a measure of the variation of the data around the mean. The more the data points differ from the average (mean) of the data, the greater the variation as measured by the standard deviation. Let's look at a few examples to help clarify the concept.

1, 1, 1, 1, 1

What is the variation in this data around its average? Zero, of course, since all of the data points are the same, and, therefore, there is no difference between each point and the average or mean of 1.

1, 2, 3, 4, 5

What is the variation in this data around its average? Well, the average is 3. And the squared difference between each data point and the average, which we will call the sum of squares (SOS), is

$$SOS = (1-3)^2 + (2-3)^2 + (3-3)^2 + (4-3)^2 + (5-3)^2$$

$$= (-2)^2 + (-1)^2 + (0)^2 + (1)^2 + (2)^2$$

$$\text{SOS} = 4 + 1 + 0 + 1 + 4 = 10$$

Now, the standard deviation is, by definition, the square root of the mean of the sum of squares. The mean is, of course, the sum of squares divided by the number of data points.

Therefore, the standard deviation (SD) for this data is

$$\text{SD} = \text{Square root } (10/5) = \sqrt{2} = 1.4142$$

Pop Quiz

Question: What is the standard deviation of the original, ten-point data set—10, 11, 11, 12, 12, 12, 12, 13, 13, 24?

Answer: The standard deviation is 3.7683.

Explanation: As we have determined earlier, the mean is 13. And, SOS = $(10-13)^2 + (11-13)^2 + (11-13)^2 + (12-13)^2 + (12-13)^2 + (12-13)^2 + (12-13)^2 + (13-13)^2 + (13-13)^2 + (24-13)^2 = 9+4+4+1+1+1+1+0+0+121 = 142$. Therefore, the standard deviation is $\sqrt{142/10} = 3.7683$.

Variance

The relationship between the standard deviation and variance of a set of data is simple. Variance is the square of the standard deviation. Conversely, standard deviation is the square root of the variance.

Hence, the statistical model for variance is:

Variance = (Standard Deviation)^2

The standard deviation of our data set, as you have just computed, is 3.7683. Therefore, the variance of the data set is 3.7683 times 3.7683, or 14.2. [What is the variance if the standard deviation is 1.4142? The answer is 2, because 1.4142 times 1.4142 is 1.99996, or 2 when rounded to the nearest integer.]

Statistical Models of Variation

Before moving on to measures of the shape of the data in a data set, let's summarize what we have learned about measures of variation or dispersion by classifying the range, standard deviation, and variance in terms of the statistical model used to compute each.

Statistical Model	Statistic
Max−Min Data Value	Range
Root of Mean of SOS	Standard Deviation
Square of SD	Variance

Measures of Shape

Do the data in a data set cluster around certain values? Or, are they uniformly spread out across their range? Alternatively, do they have some other shape, with perhaps more than one cluster of data?

Knowing how the data is distributed—the shape of the data—can be very helpful when making statistical inferences about a larger population from a sample of data. Along with measures of central tendency (mean, median, and mode) and measures of variation (range, standard deviation, and variance), measures of the shape of the data (its distribution) are an essential aspect of descriptive statistics. Together, they give us a full picture of the nature of a data set.

Once again, we will focus our investigation of data shape on the ordered-array of our original, ten-point data set:

10, 11, 11, 12, 12, 12, 12, 13, 13, 24

Earlier, we determined that the mode of this data set is 12, because it occurs more frequently than any other number. If you recall, to figure this out we listed each of the numbers in the data set, and, in parentheses, their frequency of occurrence—10 (1), 11 (2), 12 (4), 13 (2), 24 (1). If we plotted these values on a grid, they would look something like this:

4			x													
3																
2		x		x												
1	x														x	
	9	10	11	12	13	14	15	16	17	18	19	20	21	22	23	24	25

Based on this picture, what can you say about the shape of this data?

Clearly, the data clusters between 10 and 13, with one lone value, which is a possible outlier, far out to the right at 24. In fact, this value is so different from the others, that seeing it suggests the need to perform the mathematical computations (which are beyond the scope of this book) to determine if it is an outlier. Using the box-plot method, it turns out that the number 24 is a definite outlier.

This means that we can eliminate 24 from the data set on the basis that a value so far from the rest of the data is statistically extremely unlikely.

Thus, the new data set is:

10, 11, 11, 12, 12, 12, 12, 13, 13

The distribution for the new data set looks like this:

4				X		
3						
2			X		X	
1		X				
	9	10	11	12	13	14

What does this new shape tell you?

Well, the distribution is almost perfectly symmetrical. If it were symmetrical, the mean, median, and mode would be the same. [Determine the mean, median, and mode of this new, nine-point data set to convince your-self that these measures of central tendency are, in fact, quite close. Answer: Mean = 11.78, Median = Mode = 12.]

Also, this distribution looks a lot like one of the most common and useful distributions in statistics—the normal distribution. This last bit of information, if statistically verified using certain standard measures of shape (such as skewness and kurtosis), would permit us to use some very well documented methods to statistically analyze this data set. Given the central role of the normal distribution in inferential statistics, we will study it in more depth in the next chapter.

For the sake of clarification, skewness measures the degree of asymmetry in the data distribution, while kurtosis measures the degree to which the distribution is peaked or flat.

Here is an example of a data set whose distribution has a pronounced left skew. Most of the data clusters between 23 and 25, with a few of the points (skewed) far to the left.

1, 10, 23, 23, 24, 24, 24, 25

Here is an example of one with a pronounced right skew. Most of the data clusters between 1 and 3, with a few points (skewed) far to the right.

1, 2, 2, 3, 3, 3, 10, 20

Here is an example of a data set whose distribution is highly peaked. The number 12 occurs five times, which is more than twice as often as any other number in the data set.

10, 11, 11, 12, 12, 12, 12, 12, 13, 13, 14

Here is an example of one that is very flat (perfectly so, in fact). Each number appears only once.

10, 11, 12, 13, 14, 15

Pop Quiz

Question: Based on these examples, what are the skew and kurtosis of the distribution of our new, nine-point data set?

Answer: The distribution (shown in the grid on page 52) exhibits a slight left skew and a pronounced peak around the value 12.

It is important to note that there are more elaborate ways to display data distributions for analytical purposes, such as histograms and frequency polygons. However, the statistical concepts described in this chapter are fundamentals that will enable you to acquire and apply more easily the knowledge you gain later about the shape of data distributions.

Statistical Models of Shape

Before moving on to study inferential statistics in the next chapter, let's summarize what we have learned about measures of the shape of a data distribution in terms of the statistical model used to determine each.

Statistical Model	Statistic
Frequency Diagram	Frequency
Advanced Formula	Skewness
Advanced Formula	Kurtosis

Conclusion

Now you know something about descriptive statistics, the science of describing a data set using various measures of central tendency, variation, and shape.

While there is much more to learn about the subject, the material presented in this chapter will serve both an introduction to descriptive statistics and a foundation upon which to build further statistical knowledge and insight.

In the next chapter, we turn our attention to the study of inferential statistics, which is, in a nutshell, the study of statistical estimation and its pivotal role in decision making.

Chapter 5

A Primer on Inferential Statistics

Descriptive statistics answer this basic question:

"What is the nature of the data?"

Inferential statistics answer this basic question:

"What can the data tell us about the larger data set it comes from?"

Hence, if the data set is the entire population, inferential statistics are not necessary. Why?

Because descriptive statistics provide complete statistical information on the characteristics of the population, there is no need to estimate those characteristics.

If the data set is a subset or sample, however, sample statistics can help us to infer population statistics (which, by convention, are known as population parameters).

In summary, inferential statistics enable us to draw conclusions about a larger population based on a subset of the data. The most common of these inferences are:

- estimates of probability
- confidence intervals based on sampling
- hypothesis tests
- predictions based on regression analysis

We will discuss the fundamental notion behind each of these types of statistical inference in this chapter.

Overview of Inferential Statistics

Probability

Most of us have rolled the dice at one time or another when playing a board game. What if you had only one die? What is the probability that on a single roll of that die, you would roll a 2? How, in other words, do you estimate the likelihood of the next die yielding a two.

Assuming that the die was fair (not weighted or loaded to favor one side over another), you might perform the following thought experiment to figure out the answer.

Here is a list of every possible outcome from rolling one die:

1, 2, 3, 4, 5, 6

How many ways can you roll a 2? Actually, there is only one way, by rolling the die and having it land on the side with two dots.

Hence, the probability of rolling a 2 with a fair die is

$$P(2) = \frac{\text{Number of possible ways to roll a 2}}{\text{Total number of possible outcomes}}$$

$$= 1/6$$

This is your estimate of the likelihood of rolling a 2. Furthermore, it is also your estimate of the probability of any side coming up on a single roll of a fair die. [Why?]

Although the problems can get much more complicated, the notion of probability remains the same as illustrated by this simple example.

Let's try a slightly more complicated example to see if this is true. In this second case, you will roll the fair die twice in a row.

What is your estimate of the probability of rolling a total of 7 for the two rolls?

Here is a list of the possible ways that you can roll a 7. Each pair represents the number you roll on your first and second try.

(1,6), (2,5), (3,4), (4,3), (5,2), (6,1)

Based on what you learned in the previous example, what is your estimate of the likelihood of rolling a 7 in two rolls?

Well, there are six ways to roll a 7, as shown above. But, how many possible outcomes are there?

The answer is 36, which you can see from the pairs of possible outcomes that can result from rolling a 1, 2, 3, etc. shown below:

(1,1), (1,2), (1,3), (1,4), (1,5), (1,6)
(2,1), (2,2), (2,3), (2,4), (2,5), (2,6)
(3,1), (3,2), (3,3), (3,4), (3,5), (3,6)
(4,1), (4,2), (4,3), (4,4), (4,5), (4,6)
(5,1), (5,2), (5,3), (5,4), (5,5), (5,6)
(6,1), (6,2), (6,3), (6,4), (6,5), (6,6)

Hence, the probability of rolling a total of 7 on two rolls is:

$$P(7) = \frac{\text{Number of possible ways to roll a 7}}{\text{Total number of possible outcomes}}$$

$$= 6/36 = 1/6$$

Pop Quiz

Question: What is the probability of rolling a total of 3 in two rolls of a fair die?

Answer: There is one chance in 18.

Explanation: There are only two possible ways to roll a total of 3—(1,2) or (2,1). Every other possibility totals more or less than 3. [Convince yourself that this is true.] Because there are still 36 possible outcomes, $P(3) = 2/36 = 1/18$.

While, you can always use this relative frequency method to determine probability, it is not always the most efficient way. In fact, the real power of probability lies in how it represents events and in its special rules, as we shall see in the following note.

A Note on Probability

Let A be any event, like flipping a coin and having it come up heads. Then, the entire world of possibilities is contained in either A or not A, like flipping a coin and having it come up either heads or not heads (tails).

If the universe of all possible outcomes is A or not A, then what is the probability of A or not A? Simple, you say. The probability of an event whose outcome is certain, as we saw in chapter 1, is unity (1). Hence,

$$P(A) + P(\text{not } A) = 1$$

Therefore, by definition,

$$P(A) = 1 - P(\text{not } A)$$

In our simple example, $P(\text{Head}) = \frac{1}{2} = 0.5$. Therefore, $P(\text{Tail}) = 1 - P(\text{Head}) = 1 - \frac{1}{2} = 0.5$, which we know is true by the relative frequency method because the outcome of an experiment in which we flip a fair coin which does not land on its side is either a head or a tail.

A is a set or collection of elements, like flipping a head. Not A is a set or collection of elements that is the complement of A, like flipping a tail. In effect, A and not A together represent the complete universe of possible outcomes.

Pop Quiz

Question: Let A be the set of events that consist of rolling an odd number in one roll of a fair die? What is the complement of this set? What is the probability of each? Why?

Answer: Not A is the set of events that consist of rolling an even number in one roll of a fair die. And, P(Odd) = P (Even) = ½.

Explanation: A is the set 1, 3, 5. Not A is the set 2, 4, 6. Therefore, P(A) = 3/6, the number of times that A occurs versus the total number of possible outcomes. Similarly P(not A) = 3/6. Alternatively, P(not A) = 1 − P(A) = 1 − ½ = ½.

Another extremely important concept of probability theory is the notion of compound or joint events and their probability.

Let's say that two things happen in a row and you want to know the probability of the final outcome based on the probability of each thing happening. Believe it or not, this happens a lot in the real world.

For instance, what is the probability that your favorite professional team will win its division championship and go on to win the league championship? What is the probability that you will earn a scholarship to the college of your choice and will graduate? What is the probability that a computer will fail if a certain part is defective? What is the probability that a major competitor will learn of the timing of your new advertising campaign and will counter with one of its own? As you can see, the list of compound events involving a series of simple events is infinite.

Let A be the first event, and B the second event in a series of two events. (The theory extends to any number of compound events, but we will stick with two successive events to keep the illustration from becoming unnecessarily complicated.)

If A occurs first, then B occurs second, the joint or compound event AB occurs. Let A be getting into college, and B be graduating once you are in. What is the probability of AB? How do you go about estimating the likelihood of you graduating based on the likelihood of gaining admission and of graduating once you have gained admission?

Before answering this question, you need to know one more concept—the notion of conditional probability. In this case, your graduation is conditional (i.e., depends upon) a college accepting you. If you do not get into college, you cannot graduate from college.

The chain of logic for determining the probability of graduating thus goes something like this:

$$P(CG) = P(C) \cdot P(G \mid C)$$

Note that C stands for going to college and G stands for graduating.

In words, this equation says that the probability of the joint event "go to college and graduate" is the product of the probability of going to college and the probability of graduating given that you go to college.

Now that we have a formula or method for computing compound probability in this case, we need to put it to use.

What if the probability of going to college is 0.75, which means that you estimate that you have a 75% chance of admission. And, what if the probability of graduating if admitted is 2/3 or 67%. Then, you can compute the probability of the joint event that you gain admission and go on to graduate using the formula above.

$$P(CG) = 3/4 \cdot 2/3 = \frac{1}{2} = 0.5$$

Now, you have your estimate. The probability of your going to college and graduating is 0.5 or 50%.

To complete our introduction to probability theory, let's use what we learned earlier to check our estimate of the likelihood of attending and graduation from college.

Basically, we need to determine the probability (a) that you will not go to college (in which case you cannot possibly graduate) or (b) that you go but do not graduate. These events represent the complement to the event CG (i.e., going and graduating). To that end, what is the probability of not going to college? Well, if the probability of going is 0.75, by definition the probability of not going is $1 - 0.75 = 0.25$.

Second, what is the probability of going to college, but not graduating? Recognizing this a compound (i.e., joint) event, you realize that it is the product of two probabilities—of going [P(C)] and of not graduating given that you have gone to college [P(not G | C)]. Since you know the probability of going already, you need only determine the latter.

Now, P(G | C) is 2/3, so P(not G | C) must be 1/3, since you either graduate or you don't.

Finally, putting this all together you compute the probability of not graduating as:

$$P(\text{not } CG) = P(\text{not } C) + P(C) \cdot P(\text{not } G \mid C)$$
$$= 1/4 + (3/4) \cdot (1/3)$$
$$= \tfrac{1}{4} + \tfrac{1}{4} = \tfrac{1}{2} = 0.5$$
$$= 1 - P(CG).$$

Are you confused yet? If not, great! If so, you are not alone. Most students find the study of probability fascinating, but very challenging. Fortunately, the more you work with the concepts and associated formulas, the more you come to understand them.

In any event, I hope you now understand how powerful probability can be as a method of estimating (i.e., making inferences about) the likelihood of chance events, and are, as a result, motivated to learn about other forms of statistical inference, such as the estimation of population parameters based on sample statistics and confidence intervals.

Confidence Intervals

In chapter 1, we discussed sampling as a method of determining some of the characteristics of a production process. In particular, we noted that the greater the sample size, the better the estimate of the population mean based on the sample mean. As stated earlier, this finding was not a coincidence. It illustrates a very important law of statistics, called the Central Limit Theorem, which we will focus on in this section.

Let's start with a very simple example of a confidence interval. If I asked you to estimate the height of the next man who might walk by you on a crowded street, your best guess would be the average height of the typical male in your area.

Assume that the average height of the typical male is about 5′ 10″ or 70 inches. If I asked you to give me a range around that estimate within which you would felt highly confident the actual height of the next man would lie, you might say something like this: "I am 95% confident that the next man who walks by will be 5′ 2″ to 6′ 6″ in height." Thus, your 95% confidence interval would be 70 inches plus or minus 8 inches.

What does it mean to be 95% confident?

Simply put, you believe that there is only 1 chance in 20—a probability of 0.05—that you might be wrong; whereas, there is a probability of 0.95 that your confidence interval contains the actual height of the next man.

Pop Quiz

Question: What does a 99% confidence interval mean? Is it wider or narrower than a 95% confidence interval?

Answer: To be 99% confident is to believe that there is only 1 chance in 100 that your confidence interval (CI) does not contain the actual value you are estimating. Conversely, your 99% confidence interval contains the actual value with a probability of 0.99. A 99% CI is wider than a 95% CI.

Explanation: To be more confident, you need to widen the interval. For example, you might be 95% confident that the next man will be between 62 and 78 inches (i.e., 70 ± 8) in height, but you might be 99% confident that he will be 54 to 86 inches tall (i.e., 70 ± 16).

Now, what if you did not know the average height of a male in your area? What would you do then? [If you have paid attention up to this point, you will have no trouble understanding the following, multi-part answer to this question.]

Why, you would estimate it, of course. How would you do that? By picking a sample of men and measuring their height. How would you choose your sample? At random from the population in as representative a fashion as possible. Why would you pick a representative sample? To be able to use the sample mean (average height of n men) as an estimate of the population mean (the height of all men (N) in your area).

What size sample would you choose? It is impossible for you to answer this question without guessing, because we haven't covered this yet. So, let's find out how it's done so that we can answer the question of how to determine an appropriate estimate of the sample mean and, in turn, a statistically valid estimate of the population mean.

Here is where the Central Limit Theorem comes into play. As we shall see, it provides the link between the characteristics of the sample and the population.

Central Limit Theorem

A key part of the Central Limit Theorem states that if you calculate the mean (X bar) of a sample of size n, the distribution of all such means for samples of size n tends to the shape of the normal distribution as n increases.

What is a normal distribution?

The normal distribution looks like a bell. This most famous of all distributions is continuous and perfectly symmetrical around its mean. Consequently, the mean, median, and mode of the variable (X) it represents are identical.

In addition, the percentage of the area under the distribution for any interval of X is available in tables. This area represents the probability that a number lies within that interval. Because the maximum value of probability is unity (1), the area represented by any interval of X is less than or equal to 1.

The following table shows the number (K) of standard deviations (SD) on either side of the mean of a normal distribution which corresponds to a given percentage confidence interval (% CI).

CI = Mean ± K • SD		
% CI	Probability	K
50.00%	0.5000	0.68
68.26%	0.6826	1
95.00%	0.9500	**1.96**
95.44%	0.9544	2
99.00%	0.9900	**2.58**
99.74%	0.9974	3

This table says that approximately two thirds of the area of the normal distribution lies within one standard deviation on either side of the mean, over 95% lies within two standard deviations of the mean, and well over 99% lies within three standard deviations.

Consider a normal (i.e., bell-shaped) distribution with a mean of 10 and a standard deviation of 1. The 95% confidence interval for this distribution is from 8.04 to 11.96. [CI = $10 \pm (1.96) \cdot 1$]

Thus, the probability that the population mean lies within the interval 8.04 to 11.96 is 0.95.

What is the 99% confidence interval for this distribution? The answer is 7.42 to 12.58. [Prove this to yourself by following the same procedure, but with the K value that corresponds to a 99% CI.]

Now that we know what a normal distribution is, let's get back to sampling and the Central Limit Theorem.

As you will recall, this theorem states that the mean (X bar) of a sample of size n tends to the shape of the normal distribution as n increases. In fact, it is normal for $n \geq 30$ regardless of the population distribution, and is normal for much smaller values of n for most population distributions.

The Central Limit Theorem also states (a) that the mean of the sample means estimates the population mean and (b) that the variance of the distribution of sample means is the population variance divided by the sample size.

Therefore, if you select a sample of size n, you can use the mean of your sample to estimate the population mean with an error that varies directly with the population standard deviation and inversely with sample size. Because the population standard deviation is a constant, this means that by increasing sample size, you can decrease the error in your estimate of the population mean to as small a value as you wish.

This makes sense, as we discussed in chapter 1, because if you increase your sample size from n to N, the population size, the sample mean becomes the population mean and, therefore, has no error.

Before you start tearing your hair out over this statistical jargon, let's apply the Central Limit Theorem to a few simple cases and see if some of your confusion dissipates.

Case 1

You intend to draw a random sample of size n from a population with a standard deviation of 10. What must the sample size be if you want to feel 95% confident that the population mean lies within plus or minus 2 of your estimate—the sample mean?

To get a 95% confidence interval around the sample mean, we need a K of 1.96 according to the table on page 69. Therefore, K $\cdot$ SD = 1.96 $\cdot$ SD = 2. Furthermore, we know from the Central Limit Theorem that the variance of the distribution of the sampling mean is given by this formula:

$$\text{Variance of the sample mean} = \frac{\text{Population Variance}}{n}$$

Therefore, the standard deviation of the distribution of the sample mean, which is by definition the square root of the variance of the distribution of the sample mean, is:

$$\text{SD of the sample mean} = \frac{\text{Population SD}}{\sqrt{n}}$$

Hence,

$$1.96 \cdot SD = 1.96 \cdot \frac{10}{\sqrt{n}} = 2$$

Squaring both sides and solving for n yields:

n = 96.04

Therefore, you will need a sample size of at least 97 to be 95% confident that your estimate (the sample mean) is within plus or minus 2 of the actual population mean.

Let's imagine that you decide on a sample of size 100 to be on the safe side, and that the mean of this sample is 49. Then, you could state with 95% confidence that the population mean lies in the interval from 47 to 51 (i.e., 49 ± 2).

Case 2

Let's return to the earlier example and, based on what we have learned about sampling, try to estimate the average height of a man in your area.

What sample size would you choose if you wanted to be 99% sure that your estimate was within plus or minus 1 inch of the mean height of all of the men in your area?

Following the logic of the previous case, but substituting the K for 99% (which the table lists as 2.58) and 4 for the population standard deviation, you compute n:

$$n = \frac{(2.58)^2 \cdot (\text{Population SD})^2}{(1)^2} = (6.6564) \cdot (16)$$

$$= 106.5$$

Therefore, you would need a sample size of at least 107 to be 95% confident that your estimate of the average height of a male in your area was within plus or minus 1 inch of the actual population mean.

As you can imagine, there is much more to this subject. However, what you have learned should enable you to select appropriate samples and calculate related confidence intervals for most practical applications, besides providing you with a solid foundation on which to add new knowledge as you acquire it.

Hypothesis Tests

What if you invented a new product (Y) which you believed was far superior to the best brand (X) on the market. How would you prove it to skeptical consumers whose loyalty to Brand X was high?

One thing you might do would be to run a test in which you randomly assigned participants to test and control groups, and compared the responses of those who tried Brand Y (the test group) to those who tried Brand X (the control group).

After a suitable test period, you would survey both groups to see how they rated the respective products and compare the results to find out how your product,

Product Y, fared against Brand X, the best of the current alternatives.

Let Y bar be the average rating for the test group, and X bar the average rating for the control group.

Your belief before conducting the test— the alternative hypothesis—was that consumers would rate Y higher than X. Alternative to what, you say? Alternative to the null hypothesis, which is always the complement of the alternative hypothesis, and which, in this case, is that consumers would rate Y lower than or equal to X.

Note that the null hypothesis is what we assume is true unless we can find sufficient statistical evidence to refute it. In this case, the prevailing wisdom is that Brand X is superior to all. So our null hypothesis is that Product Y will receive lower consumer ratings than Brand X.

The following statements (i.e., hypotheses) summarize our assertions about the average ratings that consumers in the entire population would give each of the products if they we asked them all for their opinions about Y and X. Hence, we use population parameters in the hypotheses. Of course, our experiment will use statistical evidence based on samples, not the entire population, to test these hypotheses. [H_0 is the null hypothesis, and H_1 is the alternative hypothesis.]

H_0: Population Mean of Y $\leq$ Population Mean of X

H_1: Population Mean of Y $>$ Population Mean of X

If we had access to the entire population, we would obtain their ratings for X and Y and settle this debate right now. However, because this is not possible in our case, we have conducted an experiment involving subsets of the population divided by design into a test group, which received Product Y (the treatment), and a control group which received Brand X.

By now, you should know that even it Y bar, the sample mean of test group ratings for Product Y, is greater than X bar, the sample mean of control group ratings for Brand X, there is still a statistical probability that the actual population mean of Y is less than or equal to the actual population mean of X. So we need a statistical test of the difference between the two sample means. While the mathematics of this test is beyond the scope of this introductory book, the basic concept is not.

Let us assume that Y bar turns out to be greater than X bar. That is, the test group consumers rate Product Y higher than the control group consumers rate Brand X. So far, so good, as this suggests that our prior theory or belief (the alternative hypothesis) is correct and Y is superior to X.

To account for statistical uncertainty, we compute a 95% confidence interval around Y bar. The statistical question we want to answer is whether or not it includes X bar. If not, then we can be at least 95% confident that Product Y is superior. If, however, X bar falls in the 95% confidence interval around Y bar, then we will not be justified in rejecting the null hypothesis (that X is better) in favor of the alternative hypothesis (that Y is better).

Happily for you, you find that (a) Y bar exceeds X bar, and (b) the 95% confidence interval for Y bar does not include X bar. Hence, you reject the null hypothesis in favor of the alternative hypothesis.

Does this mean that you have proven conclusively that Product Y is superior to Brand X? Of course not, since there is still a chance that, due to sampling error, the true population mean of X exceeds or equals that of Y. But, you can say that you have demonstrated with a very high level of confidence (95%) that Product Y is better than Brand X; and that is certainly strong evidence in support of your claim.

Note that the statistical methods used to evaluate the findings of experiments can be quite a bit more involved than this example suggests. Nonetheless, as you will discover in your advanced studies, the concepts remain essentially the same, regardless of the method used.

Multiple Regression Analysis

Does leadership style influence employee performance?

Now, there's a good question; one which social science researchers have been trying to answer for some time. One of the most powerful weapons at their disposal in the quest to answer this important question is the statistical technique of multiple regression analysis (MRA).

Before delving into MRA, let's examine this question a little.

Implicit in this question is the notion that leadership style may affect how employees perform their work. That is, that performance (the dependent variable) is a result of (i.e., depends on) leadership style (the independent, controllable variable).

Expressed as a conceptual model, this theory looks something like this:

Leadership Style ⇨ Performance

Based on what you have learned about hypothesis testing, you realize that we can test this theory by specifying an alternative (and its null) hypothesis as suggested by (i.e., derived from) the theory.

Pop Quiz

Question: What are the null and alternative hypotheses for this theory about the relationship between leadership and performance?

Answer: The two, complementary hypotheses are:

H_0: Leadership style does not affect performance.
H_1: Leadership style affects performance.

Remember what I said about people using more sophisticated statistical methods to test hypotheses in real-world applications. Well, MRA is one of those methods.

In mathematical terms, MRA allows us to measure the constants A and B—called coefficients—in this mathematical representation of the relationship between the two variables in our theory—leadership style (LS) and follower performance (P):

$$P = A + B \cdot LS + Error$$

Before we proceed, it is important to point out that this equation represents the simplest of linear relationships between two variables. When applied to such a case, we refer to MRA as simple linear regression analysis.

However, statisticians normally use MRA to determine the relationship between several independent or explanatory variables (e.g., X_1, X_2 ... X_n) and the dependent variable (Y). In those cases, the mathematical model looks something like this:

$$Y = A + B \cdot X_1 + C \cdot X_2 + \ldots + N \cdot X_n + Error$$

Both the theory of MRA and the analysis of these more complicated examples are beyond the scope of our examination, but well within the bounds of most statistics courses.

That said; let us return to our analysis of the simple, two-variable model of leadership style and performance.

$$P = A + B \cdot LS + Error$$

If we postulate a continuous scale of leadership styles from autocrat to participative leader, and a continuous scale of performance from 0 to 100%, what do you think is the relationship between leadership style and performance? Specifically, in most work environments, will performance be higher or lower for leadership styles that are closer to the participative end of the spectrum than the autocratic end?

If you said higher, you're right; at least, that is what the preponderance of evidence from research shows. This makes sense if you believe that most people will work harder for someone who trusts them, allows them to participate in the decision making process, and gives them latitude in performing their work, than they will for a mistrustful boss prone to micromanaging them.

If the dependent variable (Y) is greater for higher values of the independent variable (X), there is a positive correlation (i.e., $R > 0$) between them. If Y is lower for higher X values, then there is a negative correlation (i.e., $R < 0$) between them. [What is the type of correlation between leadership style and performance? Answer: There is a positive correlation between the two variables because an increase in Y is associated with an increase in X.]

Note carefully that a strong positive or negative correlation between two variables does not mean that one causes the other. Correlation is a measure of association, not causation. Only experiments, because they provide researchers with the ability to control other variables that might influence the dependent variable, can help demonstrate the existence of a causal relationship between the independent variable (X) and the dependent variable (Y). In short, for a form of statistical proof that X causes Y, then conduct an experiment. For a statistically valid measure of the association between X and Y, then measure their correlation.

If we used multiple regression analysis to estimate the coefficients in our model of leadership and performance based on a sample of organizational members, we might obtain the following hypothetical fit:

$P = 0.1 + (0.8) \cdot LS$, where LS varies from 0 to 1.

Based on this fictitious model, performance under a pure autocratic leader (LS = 0) would be at a level of 0.10 or 10%. Whereas, performance for a pure participative leader (LS = 1) would be at the 0.90 or 90% level.

This model makes intuitive sense for several reasons. First, the greater LS, the greater P, which is e expected given the previously determined positive correlation between leadership style and performance. Second, the fact that people still work, albeit at a very low level of 10%, even for autocratic leaders, and the fact that people don't put out 100% even for highly participative leaders suggests other factors besides leadership style influence performance.

Besides these intuitive indicators of the quality of the regression equation, there are also quantitative measures, such as the coefficient of determination (R^2).

For simple models involving an independent and a dependent variable, R^2 is the square of the correlation coefficient. Correlations vary from -1 (for perfectly negative correlation), through 0 (for no correlation), to $+1$ (for perfectly positive correlation). Therefore, R^2 varies from 0 to 1. The higher the coefficient of determination, the better the fit.

Whereas the coefficient of determination is an objective measure, goodness of fit is a more subjective measure of quality. The latter examines the degree to which the regression equation, a straight line in the case of the two-variable model, matches the pattern in the data.

Notice, for instance, how closely the linear regression equation accurately represents the data on the first grid below, but not the second. The regression R^2 is high in both cases. In addition, the simple regression model—imagine a straight line through each value of x representing the regression and one through each o representing the original data—fits the first set of data perfectly.

High R2, Good Fit					High R2, Poor Fit				
4				xo	4			o	o	x
3			xo		3				x	
2		xo			2		o	x		o
1	xo				1		x			
0	1	2	3	4	0	1	2	3	4

However, in the second case a convex or dome-shaped curve that connects each value of o is a much better representation of the original data than the straight line regression model.

The morale of the story is that a high R^2 does not guarantee a good fit to the data.

There are numerous other measures of the quality of a multiple regression model that we cannot go into here. Each of these measures examines the degree to which the regression meets one of the several implicit or hidden assumptions of regression analysis. These are equivalent to, but not the same as, the criterion of randomly selected participants in the test and control groups of an experiment discussed earlier.

Pop Quiz

Question: What is the form of a simple regression model relating annual sales to advertising expenditures?

Answer: In the first case, Advertising is the independent variable, and Sales is the dependent variable. Therefore, the simple model is:

$$Sales = A + B \cdot Advertising + Error$$

If you got that one, then try this one. What is the form of a multiple regression model relating sales to advertising and price? One possible multivariate model is:

Sales = A + B•Advertising + C•Price + Error

Another possible model, which takes into account the fundamental economic principle that demand varies inversely with price (i.e., the cheaper something is, the more people will buy of it, all other things being equal), is:

Sales = A + B•Advertising + C•(1/Price) + Error

I hope you're getting the idea that MRA is an extremely practical and useful statistical method. In fact, it is one of the most important and widely used of all statistical techniques. When you get to it in your advanced studies, you will want to pay particular attention to its principles, underlying assumptions, measures of goodness, computational methods (especially templates and software written to make manual number crunching unnecessary), output formats, and real world applications.

This concludes our overview of inferential statistics. As in previous chapter on descriptive statistics, the last section of this chapter presents a table of the statistical model and relevant statistics for each inferential method covered.

Statistical Models of Inferential Statistics

Before moving on, let's summarize what we have learned about inferential statistics in terms of our generic model relating data to a statistical model to the relevant statistics.

Statistical Model	Statistical Focus
Probability Theory	Event Probability
Confidence Interval	% CI around Mean
Hypothesis Test	Reject or Accept Null
Regression Analysis	Y = f(X)

Conclusion

Now you know something about the tools of statistical analysis—descriptive and inferential statistics. In the next chapter, you will learn how to apply your new knowledge to the solution of practical problems.

Chapter 6

Statistical Decision Model in Action

Knowing the basic tools of statistical analysis is necessary, but not sufficient to enable you to solve problems in the real world. You also need to know how to integrate and apply statistical reasoning and analysis in support of decision making. After all, the ultimate goal of using statistics is to make better decisions, not to demonstrate theoretical knowledge.

In previous chapters, we focused on key pieces of the decision making process. We started with a simple statistical model (in chapter 2), which showed the direct relationship between statistics, a specific decision model, and the ultimate decision.

Statistics ⇨ Decision Model ⇨ Decision

Then (in chapter 3), we added essential information about the role of data in the statistical decision making process.

Data ⇨ Statistics ⇨ Decision Model ⇨ Decision

Finally, (in chapters 4 and 5) we introduced relevant theory about the nature of the statistical models and processes which use these data to generate the actual statistics.

Data ⇨ Statistical Model ⇨ Statistics

Now, we integrate these component parts into a coherent, five-stage model of the statistical decision making process from data collection to the final decision—the statistical decision model.

Data ⇨ Statistical Model ⇨Statistics ⇨ Decision Model ⇨ Decision

---

Substituting acronyms for two words—i.e., SM for statistical model and DM for Decision Model—results in this alternative, totally equivalent, version of the statistical decision model (SDM):

Data ⇨ SM ⇨ Statistics ⇨ DM ⇨ Decision

With this model, we can analyze practical, real-world problems, as you will see in the examples which follow.

The Backwards-Forward Process

Regardless of the problem you are trying to solve, the way you use the statistical decision model is the same. Namely, you use it backwards to plan, and forward to act—i.e., to analyze and decide.

To see how this backwards–forward process works, consider a simple situation. Imagine that you are in the market for a new TV set, and you have boiled your final decision down to set A or B. You have found a place that will give you the best buy for each set. Now all you need is some information on the quality of each set. Given this last piece of information, you can make up your mind. In other words, you have a decision model which considers cost and quality, and all you need are the respective quality statistics for sets A and B to make the decision.

What statistics do you need to make your decision? Let's say you want reliability statistics, which represent the likelihood that the TV will fail over its lifetime.

Knowing the statistics you need enables you to determine the statistical model required to generate them. In this case, you want probability based on reliability tests conducted by an independent testing agency.

With this thought in mind, you search the Internet and find a report published by a consumer rating agency that lists the very reliability statistics you are hoping to find.

In this case, you do not have to do the tests yourself, but, being knowledgeable about statistics, you read the details of the test to assure yourself that, when the agency statisticians calculated the reliability statistics, they used statistical methods appropriate to the data they had collected.

With this last step, you have completed the backwards part of the process. As a result, you now have the information you need to analyze your alternatives and make your decision.

The next part of the process takes you forward through the five stages of the statistical decision model (1) from data (which you have verified is numerical, continuous, ratio data, and, hence, suitable for the computation of reliability statistics) (2) to the statistical model (which you have determined is capable of determining accurate reliability statistics) (3) to the statistics you need (on reliability) (4) to use in your decision model (which considers cost and reliability) (5) to make the final decision between set A and set B.

As this example shows, a simple, two-way, planning and action process enables you to use the statistical decision model to determine effective solutions to practical, real-world problems. Let's try this backwards-forward process on a more complicated problem that requires us to analyze some data, and hence illustrates all ten steps of the statistical decision making process in more detail.

An Example of the SDM in Action

I once worked for a major consumer products company in marketing science, which is the study of ways to apply scientific methods to marketing. We had sophisticated marketing models that the brand managers for some of the most well known products used to make decisions about product, price, promotion, and place—the so-called 4Ps of the marketing mix.

Seeing the benefits of these models as aids to decision making, the brand managers of less well known products wanted their own models. However, because their products' revenues, while substantial, were inconsequential relative to the major brands' revenues, they were unable to get top management to invest in the development of these sophisticated marketing mix models.

When I arrived at corporate headquarters, fresh out of business school, I saw an opportunity to meet their needs by creating less complicated, but nonetheless powerful, statistical decision models based on the existence of excellent data on product variations, price, advertising, end-of-aisle in-store promotion, and distribution.

This example of SDM in action comes directly from my experience (and success) in building statistical decision models for the brand managers of popular, but less well known, national brands at this Fortune 50 Company.

As you recall, this is the basic statistical decision model (SDM):

Data ⇨ SM ⇨Statistics ⇨ DM ⇨ Decision

However, to make is easier to follow the ten-step backwards-forward process, we will use the following annotated version of the model:

Backwards to **PLAN:**

    5   4   3   2   1
Data ⇨ SM ⇨ Statistics ⇨ DM ⇨ Decision

Forward to **ACT:**

    6   7   8   9   10
Data ⇨ SM ⇨ Statistics ⇨ DM ⇨ Decision

We will apply this model and process to the creation of a statistical decision model for a single consumer product, Brand Z.

Planning Step 1—Decision

What the manager of Brand Z wants is a model that will predict the demand for (i.e. sales of) her product for the coming year based on the marketing mix variables in her control. These consist of the price of the product, and the advertising and in-store promotion budgets.

With this model, she can ask "What if" questions involving various combinations of the marketing mix variables to determine which one will maximize next year's profits from the sale of Brand Z. In terms of our statistical decision model, this optimal scenario is the desired "decision."

Planning Step 2—Decision Model

Many factors affect product demand. These include:

- Stage in the product life cycle
- Seasonality of demand
- Environmental Factors
- Marketing Mix

When a product is new to the market, its sales are typically lower than when it reaches maturity. Conversely, the sales of a product that is showing signs of becoming obsolete tend fall off from historical highs. Thus, knowing which stage of its life cycle a product is in is essential to making a good forecast of future sales. In this case, Brand Z is a mature product with stable sales.

If product sales go up at certain times of year and down at others on a regular, predictable basis, the product has an inherent seasonality in its demand pattern.

Take the demand for ice cream sold by cart vendors.
If they work in New York's Central Park, it is a safe
bet that they will sell more ice cream in the summer
than in the winter. However, if they work in the
beaches of sunny southern California, there may be
no inherent seasonality in their sales. In this case,
there is no seasonality in the demand for Brand Z.

A host of macro level factors, such as the political,
economic, social, technological, and natural climate,
also affect product demand to varying degrees. To
keep things simple, we will assume that none of these
environmental factors will change in the coming year
in ways that might significantly impact the sales of
Brand Z.

The 4 Ps of the marketing mix are product, price,
promotion, and place (i.e., distribution channels). In
this case, we assume that Brand Z will undergo no
major product or distribution changes in the coming
year. However, the sales of Brand Z have historically
been influenced by price, and two types of marketing
promotion—advertising expenditures and in-store,
end-of-aisle displays.

For Brand Z, experience has shown that sales go up
when (a) the price goes down, (b) the frequency of
television advertising insertions increases, or (c)
retail stores feature the product in an end-of-aisle
promotion. The latter consists of a temporary, sharp
reduction in price accompanied by prominent place-
ment of the product in the stores. Shoppers tend to

refer to these promotions by saying something like, "Brand Z is on sale this week."

Based on a review of the factors which theoretically affect product demand, it is clear that our statistical decision model—in this case a demand forecasting model—must account for the effects of price, advertising, and promotion on the sales of Brand Z.

Planning Step 3—Statistics

In this case, the statistics we need from our statistical model are the estimates of the coefficients in the demand model. As you know, the demand model we seek will relate annual sales of Brand Z (dependent variable) to price, advertising, and promotion (independent variables). When correctly parameterized with these statistics, it will enable the manager of Brands Z to ask and answer "What if" questions and, thereby, determine the optimal marketing mix.

Planning Step 4—Statistical Model

From our discussion in the previous chapter, you know that one possible multivariate statistical model involving the three independent variables—price (P), Advertising (Ad), and in-store promotion (Pr)—and the dependent variable—Sales—is

$$Sales = A + B \cdot P + C \cdot Ad + D \cdot Pr + Error$$

Another possible multiple regression model which relates these variables is captured in this equation:

$$Sales = A + B \cdot (1/P) + C \cdot Ad + D \cdot Pr + Error$$

In practice, we could try either one and see which fits the data better, but this would be bad practice. Why?

Because, statistically the tests applied to determine the quality of the regression are tests of the null hypotheses that the overall regression and the coefficients of each of its terms—A, B, C, and D in our case—are zero. As you know, you get only one chance to test any hypothesis, not as many as you want. Otherwise, the statistics you generate may be bogus, or as statisticians say, spurious. Spurious significance is something that you want to avoid.

So, which should we pick as the basic model? If you believe or have data that shows that, all other things being equal, price and sales vary in a more or less linear fashion over a wide range of prices and associated sales, then the first model is fine. However, because we have strong prior economic theory that price and sales are, in general, inversely related, we should go with the second equation as our basic regression model.

Planning Step 5—Data

To estimate the coefficients in a multiple regression model, numerical interval or ratio data is necessary.

What type of data do we have in this case?

Sales data (i.e., product demand data) is numerical continuous ratio data (e.g., 5,675,347 units sold per year). So are annual advertising expenditure (Ad) and price (Pr). [Convince yourself of this before moving on.]

So far, so good; but, what type of data is in-store promotion data? Now, that's a tricky question. The answer depends on what an in-store promotion is.

For the sake of simplicity, let's assume (a) that our data is monthly data, (b) that all in-store promotions, when they run, run for the same time period, such as a week, during any given month, (c) that when they run, they run in all stores, and, finally, (d) that only one promotion runs in any given month.

As a result, we can use a dummy variable to represent the presence or absence of a national, in-store promotion for Brand Z in a given month. A dummy variable is either 1 or 0 depending on whether a promotion occurred in that month or not. In short, if there was a promotion, then the promotion dummy variable has a value of 1. If not, the dummy variable has a value of 0.

Based on this operational definition of an in-store promotion, what type of data will we create to represent promotions? The answer is numerical discrete ratio data (i.e., the dummy variables for each month for in-store promotion will have either the value 0 or 1).

Having used the steps implicit in the statistical decision model in reverse order to figure out how to enable the Brand Z manager to make the decision about the optimal marketing mix for the coming year, we now use these same steps in their normal order to create the decision model and apply it to make this decision.

Action Step 6—Data

If this were a real project, you would collect data monthly for each of the four variables in the regression model, or obtain and verify the accuracy of existing data from the company's archives.

Let's imagine that you have done so, and that the table below displays a small portion of the data for a six month period during the past several years:

Sales	Price	Advertising	Promotion
114	40	90	1
103	40	120	0
126	40	120	1
107	50	140	0
122	50	130	1
111	50	150	0

Note that, for the sake of not getting unnecessarily confused by fictitious numbers, we express the real sales (and advertising) data as indexes based on average historical sales over the time interval represented by the data. Thus, sales of 115 indicate that in that particular month sales were 115% of normal (i.e., 15% above the average). Likewise, advertising of 110 indicate that advertising expenditures were 110 % of the historical average in the given month.

As discussed earlier, a zero (0) for promotion means that no in-store promotions occurred in that month; whereas, a one (1) means that a promotion occurred.

Just as an exercise in the power of statistics (in the form of multiple regression analysis in this instance), examine the data in the table and try to figure what portion of the monthly variation in sales is due to changes in price, advertising, or promotion.

Action Step 7—Statistical Model

To keep things simple, even though the data in the table is only a tiny fraction of the entire sample of several years of monthly data, I used it to compute the coefficients of the following regression equation:

$$\text{Sales}_{rel} = 26.33 + 1200 \cdot (1/P) + 0.4 \cdot \text{Ad}_{rel} + 21 \cdot \text{Pr}$$

The sales index computed from this equation (Sales_{rel}) is an estimate of relative monthly sales. To obtain an estimate of monthly sales, we simply multiply this estimate of relative sales by the average historical sales used to create the indexed data.

Action Step 8—Statistics

To compute the statistic—estimated sales—for each combination of marketing mix variables chosen by the Brand Z manager, it is only necessary to substitute the values in the statistical decision model above. As an example, consider a price of 50, advertising pressure (Ad_{rel}) (which is an index of advertising expenditures relative to the historical average) of 100, and a promotion dummy variable value of 1. Using the regression model, as shown below, we compute the estimated relative sales:

$$Sales_{rel} = 26.33 + 1200 \cdot (1/50) + 0.4 \cdot 100 + 21 \cdot 1$$

$$= 26.33 + 24 + 40 + 21 = 111.33$$

If average historical sales are 1 million units sold per month, then the estimated monthly sales for this combination of price, advertising, and promotion are 1.1133 million units.

Perhaps you noticed that the statistical model estimates monthly sales, not annual sales. Yet, the stated objective was to estimate annual sales. The main reason for this is that, as you may recall, the company either runs or does not run in-store promotions once a month. In addition, brand managers often vary the timing and size of advertising pressure (i.e., media expenditures relative to historical average expenditures) for a variety of competitive and cost reasons. Hence, it is better to develop a monthly sales model, run it based on a twelve month marketing mix scenario, and cumulate the results to obtain an annual sales estimate for the Brand Z manager.

Before moving on to discuss the decision model, it is important to stop and examine the quality of multiple regression model to see if it measures up. Because the statistical model uses only a small, and potentially unrepresentative, subset of the entire multi-year data sample, it would be very unwise to make any real predictions based on the model as fit. Nevertheless, if we assume for a moment that it is representative of the larger sample, we can examine its quality as a statistical model.

You may recall that R^2, the coefficient of determination, is one measure of regression quality. The closer to 1 it is, the better the regression in the sense that it explains more of the variation in the data around its mean or average value. For this model, the R^2 is an excellent 0.9862, or almost 1.

In addition, many of the other measures of quality that are beyond this treatment of the subject are quite good, including tests of the statistical significance of the overall regression and its individual coefficients. All is not perfect, however. There are indications of a fairly high degree of correlation between the independent variables, which can cause serious problems if not addressed. Nevertheless, the statistical model is sufficiently robust, assuming as we have that the data used to fit it are representative, to use its outputs in the decision model.

Action Step 9—Decision Model

The decision of what the marketing mix variables should be at any time for a given product depends on the impact that decision has on bottom-line profits. High sales are fine, but if achieved as a result of even higher cost, they are a prescription for long run failure.

This is why the decision model for Brand Z considers sales and the cost of generating those sales. To keep matters simple, we assume that the brand managers in this particular company have a total budget which they can allocate to either advertising or promotion, or some combination of the two. Because the Brand Z manager will want to use every dollar in her budget to market her product, the cost constraint is, therefore, a constant for all combinations of advertising and promotion.

If the price of the product (Brand Z) was not subject to change in the coming year, then finding the combination of marketing mix variables that maximize sales would also optimize profit. However, changes in price affect not only product demand, but also revenue. This is because sales revenue is the product of sales (in units) multiplied by price (per unit). Expressed as a formula, sales revenue is:

Sales Revenue = Unit Sales • Price

= function(1/Price) • Price

Thus, if the Brand Z manager increases price, unit sales, which are an inverse function of price, will go down. The question of whether sales revenue will go up or down as a result of the price increase depends on the price elasticity of demand. Fortunately, the regression model, our statistical model, incorporates the effect of price elasticity along with the effects of advertising and promotion.

The point of this discussion is that, in a real-world application, the statistical model is typically not the decision model itself, but one input to a more complicated model that incorporates other factors relevant to the final decision.

As we shall see in chapter 7, the art and science of building decision models is not in the domain of statistics, but falls in the province of an entirely different discipline, which some practitioners refer to as Management Science and others call Operations Research.

Action Step 10—Decision

The decision model is now complete and ready for the Brand Z manager to use. She will develop a series of feasible scenarios involving various combinations of marketing mix variables, and use the decision model to forecast sales revenue and profit for each. The result of this "What if" exercise will be the optimal marketing mix based on the model. She will then incorporate the effect of other, non-quantifiable factors, to arrive at a final decision about the best marketing mix for Brand Z in the coming year.

Conclusion

At the beginning of this chapter, we set out to integrate the material covered earlier in the book into a coherent model, which we called the statistical decision model, and to apply that model to a reasonable facsimile of a real-world problem.

Applying the model (reproduced below), to the problem of forecasting the annual demand for a mature product by using relevant marketing mix variables enabled us to explore the ten steps of the backwards-forward process in some detail.

Data ⇨ SM ⇨ Statistics ⇨ DM ⇨ Decision

As a result, you should now have a much fuller and more profound appreciation of the power of statistical reasoning, analysis, and tools to improve the quality of real-world decision making.

Chapter 7

What's Next?

When we started, I said that this book was for problem solvers and decision makers, not mathematicians. I also said that it was an integrated treatment of the subject of statistics, leading progressively from fundamental concepts to more advanced ideas.

I hope I have delivered on these promises, because now your real work begins. For most of you, the next step will take you straight into a traditional statistics course, where theory predominates, applications are greatly simplified, and quizzes do count in the final grade.

I sincerely hope that the time we have spent together examining the fundamentals of statistical reasoning and analysis will pay dividends by providing you with a big picture view (a road map so to speak) and, thereby, preventing you from getting lost in the myriad of essential details that mastery of an important subject like statistics will require you to study.

I also hope that you will discover how much of the theory of statistics you have learned, somewhat less painlessly I imagine than from a traditional, theoretically-focused book, as a result of this book's steadfast focus on the practical application of statistics to real-world decision making.

Finally, I hope that I have peaked your curiosity about decision models and decision making by showing you how data, statistical models based on them, and statistics generated by those models can provide valuable inputs to these larger, more all-encompassing decision tools.

If I have, I strongly urge you to investigate the field of Operations Research/Management Science. This is the true domain of the quantitative problem solver—a person who attempts to harness the awesome power of mathematics and science to solve complex, practical, real-world problems on a daily basis.

As someone trained in this area who applied it to a host of fascinating technical and management problems during a successful career in the field, I assure you that this avenue is one that is educationally, experientially, and financially well worth pursuing.

Whether you are content to use this book to ease your transition into a traditional statistics course, which is the primary reason I wrote it, or you decide to pursue the broader domain of quantitative problem solving that includes, but is not limited to, statistics, is, of course, up to you.

Either way, I wish you well.

Dr. L

References

Berenson, M. L., Levine, D. M., & Krehbiel, T. C.
(2004). *Basic business statistics: Concepts and
applications* (9th ed.). Upper Saddle River, New
Jersey: Pearson-Prentice Hall.

Marrow, A. J. (1969). *The practical theorist: The life
and work of Kurt Lewin.* New York: Basic Books.

McGregor, D. (1960). *The human side of enterprise.*
New York: McGraw-Hill.

Mish, F. C. et al. (1997). *The Merriam-Webster dictionary.* Springfield, Massachusetts: Merriam-Webster,
Incorporated.

About the Author

Robert E. Levasseur, Ph.D., a full-time faculty member at one of America's premier online Ph.D. granting universities, teaches doctoral courses and mentors Ph.D. students in Leadership and Organizational Change, Information Systems Management, Operations Research, Engineering Management, Accounting, and Knowledge Management.

Dr. Levasseur earned undergraduate degrees in physics and electrical engineering from Bowdoin College and MIT, and master's degrees in electrical engineering and management from Northeastern University and the MIT Sloan School of Management. His Ph.D. is from Walden University.

Dr. Levasseur has taught for Boston University, Anne Arundel Community College, and the University of Maryland University College part-time; and for the University of the Virgin Islands and Walden University full-time.

Dr. Levasseur's professional career spans over three decades and includes leadership, management, and organizational change positions in Fortune 50 corporations. He is a registered Organization Development consultant and a member of INFORMS, the Institute for Operations Research and the Management Sciences.

Dr. Levasseur is the author of numerous articles and books. These include *Breakthrough Business Meetings,*

Leadership and Change in the 21st Century, Practical Statistics, and *Student to Scholar*.

A native of Sanford, Maine, Dr. Levasseur and his wife live on the shores of the Chesapeake Bay in Annapolis, Maryland. To learn more about "Dr. L" and his work, visit his web site at www.mindfirepress.com.

Books by Robert E. Levasseur

Breakthrough Business Meetings

Leadership and Change in the 21st Century

Practical Statistics

Student to Scholar